A Maker's Guide to

PULLEYS

Written by JOHN WOOD

Illustrated by AMY LI

LERNER PUBLICATIONS ◆ MINNEAPOLIS

First American edition published in 2021 by Lerner Publishing Group, Inc.

Editor: Madeline Tyler
Design: Drue Rintoul

Lerner Publications Company
An imprint of Lerner Publishing Group, Inc.
241 First Avenue North
Minneapolis, MN 55401 USA

For reading levels and more information, look up this title at www.lernerbooks.com.

Main body text set in VAG Rounded Std. Typeface provided by Adobe Systems.

Library of Congress Cataloging-in-Publication Data

Names: Wood, John, 1990- author. | Li, Amy, illustrator.
Title: A maker's guide to pulleys / written by John Wood ; illustrated by Amy Li.
Other titles: Pulleys
Description: First American edition. | Minneapolis : Lerner Publications, 2021. | Series: A maker's guide to simple machines | Includes index. | Audience: Ages 6–10 | Audience: Grades 2–3 | Summary: "Inquisitive makers will love learning about pulleys and how they work thanks to the Maker, a wacky alien with a passion for simple machines! A hands-on back matter activity adds to the fun"— Provided by publisher.
Identifiers: LCCN 2020054217 (print) | LCCN 2020054218 (ebook) | ISBN 9781728416434 (lib. bdg.) | ISBN 9781728438320 (pbk.) | ISBN 9781728418889 (eb pdf)
Subjects: LCSH: Pulleys—Juvenile literature.
Classification: LCC TJ1103 .W66 2021 (print) | LCC TJ1103 (ebook) | DDC 621.8—dc23

LC record available at https://lccn.loc.gov/2020054217
LC ebook record available at https://lccn.loc.gov/2020054218

Manufactured in the United States of America
1-48890-49204-12/17/2020

Table of CONTENTS

Words that look like this can be found in the glossary on page 24.

BE A MAKER

Oh no! Sorry, Maker has just eaten breakfast. He is a very messy eater. Say sorry, Maker.

Believe it or not, Maker is very smart when it comes to machines.

A machine is an object that makes a job easier to do. Maker wants to teach you about one of the simplest types of machine: a pulley.

Scissors are a kind of machine. So is a bicycle.

PULLEYS

Pulleys are a type of simple machine that are used to lift heavy things. There are lots of different types of pulley systems for lots of different jobs. Each one looks a little bit different.

All pulleys have at least one wheel and a rope. The wheels of a pulley usually have grooves in them. This is so ropes easily fit around them.

Pulleys can use wire, string, chains, or even steel cables instead of rope.

Here are some examples of pulleys.

Cranes use pulleys.

Pulleys are often found on boats and ships.

Pulleys are used to raise flags.

This well's bucket is lowered on a pulley.

Maker says that there may even be pulleys in your home. Here are a few.

Some clotheslines use pulleys.

Some garage doors open using pulleys.

Pulleys move blinds up and down.

PARTS OF A PULLEY

Pulley wheels with grooves like this one are called sheaves.

An axle is a small rod that goes through the middle of a wheel. The axle lets the wheel spin. Sometimes the axle is fixed to the wheel, and sometimes the wheel spins freely around the axle.

The weight is the thing that you want to lift.

HOW A PULLEY WORKS

There is a part of a pulley that pulls one end of the rope. The other end of the rope is tied to an object. The heavier the object, the more force is needed to lift it.

This factory worker is controlling a crane with a remote control. The remote control is connected to a motor.

A pulley's ropes can be pulled by humans or a motor.

Friction is a force that happens when two things rub together. Friction slows things down and makes heat. There is friction between the rope and the pulley wheels.

You can feel the heat from friction when you rub your hands together quickly.

It takes a little bit of extra force to overcome friction.

FIXED OR FREE?

Pulleys can either be fixed or free. In a fixed pulley, the wheel stays in the same place, and only the rope, wire, or chain moves.

A flagpole is an example of a fixed pulley.

A free pulley has a wheel that moves. A zipline is an example of a free pulley. The wheel moves along the wire.

A pulley system can be a mixture of fixed and free, where some wheels move and others don't.

MORE WHEELS, MORE POWER

When a pulley has more than one wheel, it becomes easier to lift the object. If a pulley has two wheels, it takes half the force to lift the object.

Pulleys help people lift objects that would be too heavy to lift on their own.

More wheels mean even less force is needed. This is because the weight is evenly spread over the different parts of the rope. This is more efficient than having all the weight in one place.

Four wheels will take a quarter of the force to lift the same object as one wheel.

BUILD AN ELEVATOR

It is time to build! We will be using a pulley to make an elevator that moves up and down. Have an adult help you use scissors.

YOU WILL NEED:

- A tall, rectangular cardboard box
- A smaller, square cardboard box that will fit inside the rectangular box
- Scissors
- A long piece of string
- A straw
- A small, thin stick
- A pencil
- A piece of cardboard
- Some colored pencils

STEP 1

Cut the front off both cardboard boxes. Trim the straw so it is half the width of the rectangular box or shorter.

STEP 2

Push your stick through one side of the cardboard box, near the top. Slide the straw over the stick.

STEP 3

Push the stick through the opposite side of the box. Keep the stick level, like it is in the picture.

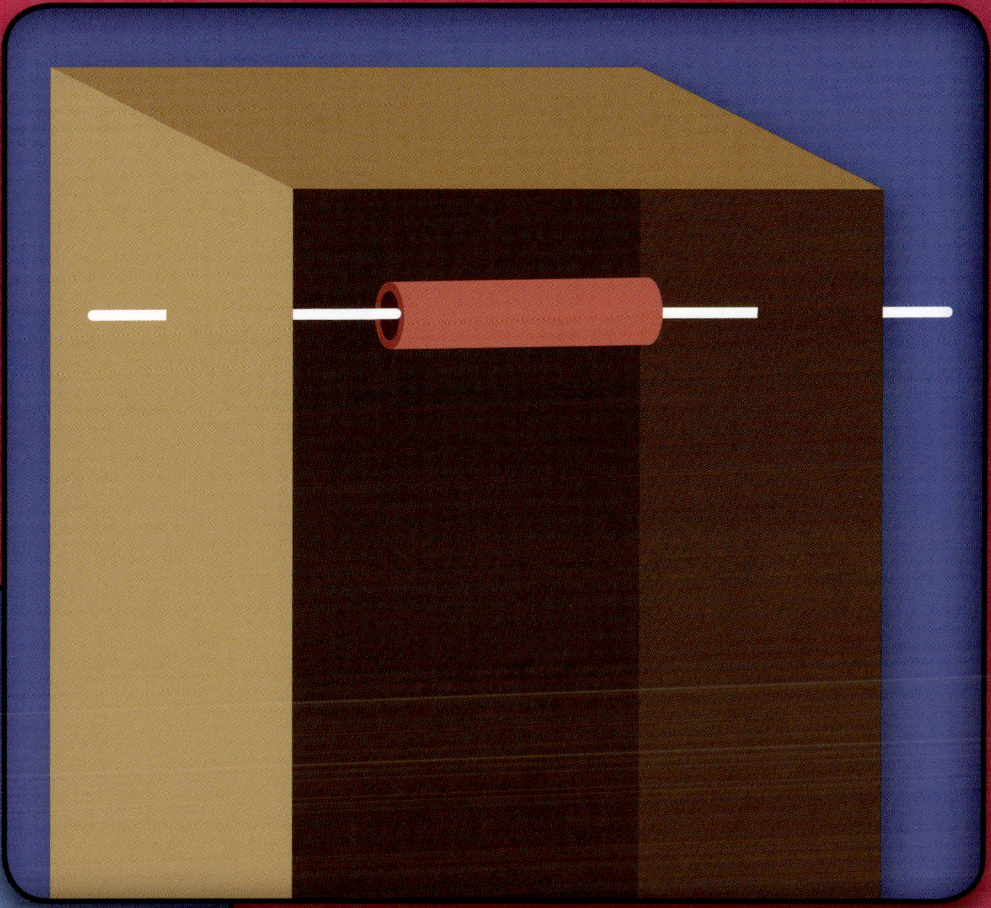

STEP 4

Make a hole in the back of the rectangular box, near the top. Now make a hole in the top of the smaller box.

STEP 5

Push the string through the hole in the smaller box and tie a knot at the end.

STEP 6

Put the smaller box inside the rectangular box. Loop the other end of the string over the straw, and push it through the hole in the back of the rectangular box.

STEP 7

Use your pencils, scissors, and piece of cardboard to make a character.

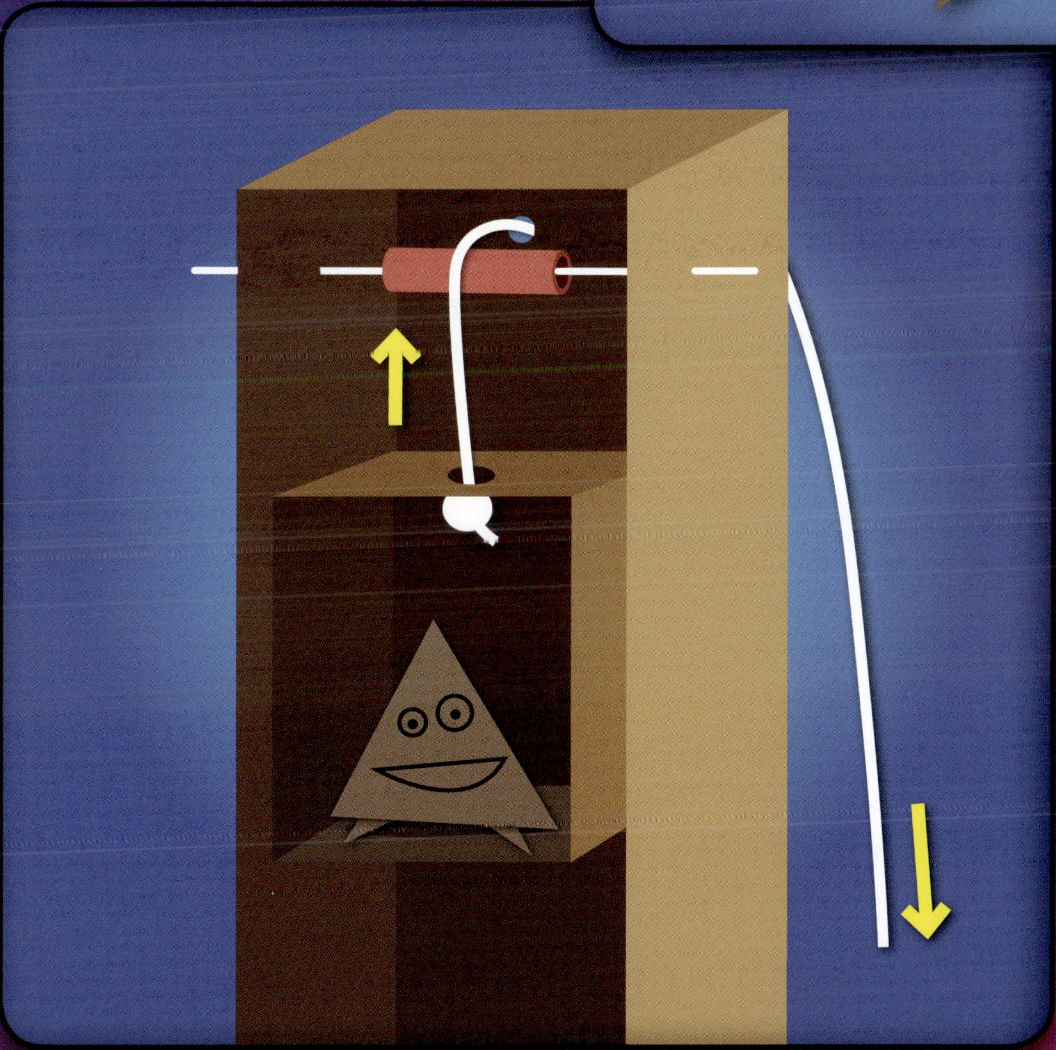

STEP 8

Put your character in the elevator. Grab the end of the string behind the box and pull it downward to move your elevator!

GLOSSARY

force	a push or pull on an object
grooves	cuts or marks made in hard objects
motor	machines that make things move
overcome	to defeat, overpower, or deal with something
systems	sets of things that work together to do specific jobs

INDEX